SHATTERED

By

Caleb Ekeh

For permission requests, write to the author, addressed to:

Caleb Ekeh

For Erin Wing, a great teacher, and supporter of my writing.

"Good morning, Mr. Kavrich. It is July 12th, 2075," an automated voice repeated.

Elijah Kavrich slowly opened his eyes, the white fluorescent light of the laboratory blinding him as he lifted his head from his desk.

"Another late night," Elijah grumbled, massaging his temples.

"I gather it is true what they say, genius never sleeps," the automated voice said.

"I don't think I programmed you with a sense of humor, Rhombulus."

"That you didn't, sir. Perhaps my circuits have just become fried with anticipation for the great day," the automated voice mimicked a chuckle as its vessel, a small white three-dimensional rhombus, floated down from the ceiling, held in the air by advanced thrusters.

"Yes, today is the day we change the world," Elijah smiled as he straightened his lab coat.

"Sir, I am a mere artificial assistant. You will be the one to change the world," Rhombulus said as his singular blue optic lens examined his creator.

"Remember, it took two people to invent the plane. Wilbur was just as helpful as Orville." Elijah turned his attention to the digital hologram in the center of the room.

"You have triple-checked everything, sir. The calculations are flawless... this machine will revolutionize energy," Rhombulus responded, sensing doubt.

"This is what I get for giving an A.I. emotional receptors," Elijah chided. "But I guess it's too late to make any changes to the project. It's down at the plaza by now." He turned his attention to the door.

"I already have breakfast prepared for you, a bowl of oats, your favorite," Rhombulus said, following close behind Elijah.

Elijah walked through the nearly empty lab building. The halls were lined with holographic displays of the company's previous inventions. He came to a stop at the most prominent display.

"In 2045, a century after the first atomic bomb laid waste to Hiroshima, Alexandrius Heptner, Founder of Atlas Energy Corporation, designed the Advanced Molecular Pylon (AMP), a nuclear generator capable of generating more electricity than any other, and the most compact. It revolutionized the field of energy and made Atlas Energy the behemoth of the industry it is today. We owe our thriving metropolis of New Athens to Mr. Heptner. Atlas Energy: We support the world…" The recording finished.

"Truly an inspiration," Elijah marveled as he envisioned his own invention displayed in these hallowed halls.

Elijah moved to the break room, where a bowl of oat cereal had been prepared. He sat down to eat.

"Have I received any messages from anyone?" he said with a mouthful of cereal.

"You have received one message from Lilith Kavrich. I will play it for you now," Rhombulus said.

"I am getting tired of our double bed being single... But I know you are busy. Good morning, hun. I can't wait to see you at the unveiling today. Molly can't wait either. The poor girl fell asleep waiting for you to come home last night. Love you." The message ended.

Elijah smiled as he finished his oats.

"It seems you have quite a fan club, sir," Rhombulus said.

"Yes, probably shouldn't keep them waiting." Elijah stood and went outside the complex.

"I await your return, sir. I will prepare the balloons and streamers," Rhombulus said, stopping at the door.

The metro had begun to pull into the station. The white train gleamed in the morning sun as its doors opened to its passengers.

"This train will now depart for Founders Plaza. Please hold tight to the railing as we depart," the train's automated voice announced as Elijah entered the train.

The soundless train began to pick up speed, moving through the city on mechanical rails suspended high in the air. It weaved through the tall skyscrapers, and office workers waved to the passengers from their cubicles. The train's rails descended as they approached the large lake in the city's center, bringing them close to the water. Founders Plaza was located on an island in the center of the lake. The waters were clear as glass and made great drinking water, thanks to purifying technology.

The train stopped gently in the plaza—a diamond-shaped platform of smooth stone with a spire in the center. The spire was an elegant technological fountain that represented the peace and prosperity of New Athens, sustained by technology. Large crowds of people had gathered to witness the unveiling of Elijah's new creation.

Elijah walked onto the stage where his creation remained hidden beneath a cloth.

"The presentation will begin within the hour," Francine Glebor, the CEO of Atlas Energy, said as she shook Elijah's hand.

Elijah searched the crowd for his family and found them near the front row of people.

"Hello, my little Molecule!" Elijah ran to his daughter, picking her up.

Molly Kavrich laughed as she hugged her father.

Lilith wrapped an arm around them both.

"Yes, she dragged me out of bed bright in the morning so we could get a good spot," Lilith smiled at her husband.

The family stayed there for the next hour. In the presence of his family, Elijah's words were different; he did not talk about science or technological advancement, but rather memories of past outings and future plans for fun.

"The Annual Tech Demonstration will begin in 10 minutes," a robotic announcement fluttered through the crowd.

"I guess I should get up there," Elijah said, kissing his daughter's forehead again.

"I have made reservations for dinner this evening. I look forward to celebrating your success," Lilith said, giving him a parting smile as he turned to leave.

Elijah took the stage, standing with several of his colleagues. He watched as Francine approached the podium.

"Welcome, everyone, to the 40th Annual Tech Demonstration, where we celebrate another year of technological advancement," Francine smiled as her attention turned from the audience to the camera drones that circled overhead. "It is my pleasure to start this event with a demonstration from someone from my company... Elijah Kavrich." Francine stepped aside.

Elijah took the podium. "Thank you, Mrs. Glebor. I am pleased to introduce to you all... the future of energy," Elijah said as the nano-cloth behind him faded away, revealing a large machine. "The gravity generator! Capable of generating near-infinite energy thanks to the power of gravity!" Elijah smiled as he listened to the applause of the audience.

The machine's steel plating shimmered in the sun. A structure standing 20 feet tall, consisting of a large weight at the top held by a magnetic lock, and a large pressure pad at the tower’s base.

"With this key, I turn on this generator and power our future," Elijah said as he turned the key.

The audience watched in amazement as the weight was released from its lock, and it seemed the whole world went silent. The weight slowly came down like a mighty hammer, hitting the pressure plate with immense force.

But something unexpected happened.

A large shockwave was generated, and everyone was sent floating into the air as if gravity were nonexistent.

"Something is wrong..." Elijah said to himself as he floated in the air.

His thoughts were disrupted when the weight came down again, sending another shockwave that caused everyone in the air to crash back down to the ground. Screams were heard as this cycle continued several more times.

"Someone turn it off!" a crowd member yelled as a chorus of screams rose from the audience.

The machine had gone into overdrive now, the weight moving like a heavy piston, *up and down, up and down,* each time striking the pressure plate with immense force. A loud cracking noise filled the air as large cracks began to form in the ground.

Elijah ran to his family, wrapping his arms around them in protection.

"I... don't know what's happening... I ran the calculations," Elijah said as he looked at the surrounding chaos. The crowd had begun to scatter, running in all different directions, trying to get as far away from the machine as possible. The cracks in the ground had grown wider, and many people fell into the opening chasms.

"I caused this..." Elijah's words ran hollow over the chaos as the ground beneath him crumbled.

***

"BEEP!" A loud ringing sound filled the darkness. Ely slowly opened his eyes as he listened to the various pieces of technology implanted in his chest come to life.

"A20 OXYGEN SYSTEM ONLINE," the machine said.

Ely breathed in the toxic air around him and could hear the sound of the machines in his lungs working to purify it.

"H06 HEART MONITOR ACTIVE," another machine said.

Ely stood from his bed and felt an electric shock run through his body as the heart monitor increased his heartbeat.

He cursed. "Stupid machine." He banged on his chest as he walked around his small apartment.

The slum had been the only affordable option, and it stunk of rat and human waste. He opened the blinds and looked out at his view. Beyond his window was a barren wasteland, the outside of the city. The surrounding area had become arid due to nuclear fallout and was barren gray rock as far as the eye could see. In the distance loomed a large chasm.

"Shatter scars," a voice said from behind him.

"Looks like I'm not the only one awake," Ely said as he turned to look back at his roommate.

"Shatter scars... reminders of the Shattering, which happened 50 years ago. It is a reminder that the 16 chunks or so of our world are barely holding together," he continued, ignoring Ely's remark.

"Are you always a walking encyclopedia, Barry?" Ely said.

"No, but do you always have something snarky to say?" Barry responded sharply.

"Is the kid awake?" Ely's face became grim.

"No, he has been out for quite a while," Barry responded as the two moved to the corner of the room where a small dog bed had been made up with blankets and pillows. Tucked comfortably in the dog bed was a small boy.

"How is he doing?" Ely ran a nervous hand through his greasy black hair.

"Well, his heart monitor is out of juice, meaning he will likely die within a day or two." Barry gently took the boy's hand.

"Is that you? Uncle Barr, Uncle Ely?" the boy stirred.

"Yeah, we're here, Redge," Ely smiled, taking the boy's other hand.

"I'm hungry," Redgie clutched his stomach.

"Get the cup noodles... make 'em up real nice," Ely said, turning to Barry.

After a while, Barry returned with three Styrofoam cups of noodles.

"Yum yum in my tum," Ely said as he spooned a clump of noodles.

Redgie laughed, which made all three of them smile.

"Hey... Ely?" Redgie looked at him. "Barry sometimes tells me stories about things from before... Is it true that families used to sit and eat good meals at a table?"

"Barry needs to learn to keep his yapper shut," Ely glared at him. "No... that's just stuff from stories now, y'know? It ain't real... not today."

"Oh... that's too bad. What happened to them?" Redgie asked, his eyes filled with curiosity.

Ely sighed, thinking carefully about his response. "Welp, what I've heard is just the same as Barry, from old news articles and old-timers who rant and rave, but apparently, after the Shattering, people lost interest in speaking to each other. They turned to their pocket phones and TV screens. Technology was their first mistress, per se. It could bring them joy, comfort, power... even pleasure on some occasion," Ely said.

"I don't understand."

"Me either, kid," Ely sighed.

"You think we will ever eat dinner together like that?" Redgie asked innocently.

"Me and Barry need to have a little uncle chat." Ely abruptly stood and dragged him into the next room. "What were you thinking... telling him Old World junk!" Ely yelled.

"I thought it would cheer the kid up!" Barry said, raising his own voice.

"No need to put those kinds of ideas in the kid's head, Barry!" Ely abruptly raised his hands.

Barry cursed. "We both know the kid isn't going to live long enough to even put those ideas into fruition!"

Both men went silent as Barry's words lingered in the air.

"Look, I—" Barry began to say something.

"Shut it!" Ely shouted as an angry sob escaped his lips.

"Ely—" Barry began to speak again until he was interrupted by a glass shattering. Ely had thrown a glass at the wall.

"I ain't going to let that kid die, Barr," Ely said, desperation in his voice.

"What... you suggesting then?" Barry said.

Ely opened the drawer and pulled out a black pistol. "I'm saying we'll steal the kid a new heart."

"You're crazy. Done... lost your mind." Barry walked away, throwing his hands up.

"Barry, listen to me—"

"No! What happened to reforming? That's not us anymore."

"We're doing this for the kid, though," Ely said as he began to polish the gun.

"You know what the Imperial Forces do to people who rob organ implants... It's not pretty, let me tell you," Barry said.

"In and out. We won't get caught. And if we do..." Ely cocked the gun.

Barry went silent before complying with a sigh. "I'll start the truck."

"Atta boy, Barry." Ely began to tuck the gun into his coat pocket. Before walking to the door, he stopped over Redgie's sleeping form. "Get some rest, kid. We'll be back in a jiff," he smiled as he gently ruffled Redgie's hair.

The beige-colored pickup truck was waiting in the driveway. Unlike every other vehicle in the city, the relic truck of the Old World still required a set of physical keys to function.

Barry inserted the keys into the cylinder and watched the engine sputter to life.

"It may be bad for the environment, but it's good for the soul," he laughed as they sped down the street in a cloud of dust. They traveled several miles until they reached the central city, where an armed checkpoint awaited.

A tall man in jet-black combat armor approached, holding a rifle.

"Identification?" he asked as the blue reticles of his closed-in helmet landed on the two.

"One minute," Barry said as he rolled up the window. "Stupid Imperial Forces and their identification," he said as he began to rummage through the jockey box of the truck, saying every curse word known to man.

"Ay, just 'cause the kid isn't here doesn't mean you can use your full vocabulary." Ely slapped the back of his head.

Barry gave him a confused look.

"Don't give me that, you know what I mean, idiot," Ely said as he began using his own collection of swear words. The two began to argue back and forth, their voices becoming yells and shouts.

Their argument was interrupted by a knock on the window.

Barry rolled down the window. "Here is our identification," he said, handing the guard two cards.

"Proceed ahead," the guard said as he signaled for the security gate to open.

The two watched as the large steel gate shuddered before opening wide. They drove into the city.

The Shattered City was a maze of gleaming skyscrapers and flashy billboards. Most people in the streets wore the finest clothing and were equipped with multiple tech implants. Technology made their bodies more appealing, giving their eyes a fresh glow, refining their muscles, and removing all skin defects.

"These plastic folk always give me the creeps," Barry said as he shivered.

"Don't worry, our destination is just up ahead," Ely pointed to the building down the street.

The building was near the city's center and was one of the tallest buildings. It had hundreds of elegant glass windows and was made of fine steel. Out in the courtyard of the building, a holographic sign read "The Corporation."

"Ah yes, the greedy conglomerate that makes everything from high-tech implants to the very food we eat," Barry remarked, his voice laced with disdain.

"Like I said… in and out," Ely reassured.

They parked the car in an alleyway next to the building.

"So how are we going to play this?" Ely asked.

"I was thinking of a simple diversion: we go in and ask for a tour of the facilities, pretending we're some interested tech scholars, I don't know. Once we discover where they keep the heart monitors, you break off, break in, and steal the loot, and we meet back outside," Barry said.

"Solid plan. I like it," Ely nodded his head.

"Here, take this. It will deactivate any locked doors or security systems," Barry handed him a thumb drive.

The two made their way into the building lobby.

"Hello, welcome to the Corporation. How may I be of assistance?" the woman at the front desk said.

"We're out-of-towners, you see, and we are interested in a little tour of the facilities, just to see how things are," Barry said.

"Yes, of course! I will prepare a tour for you immediately." The lady moved to the computer. Within several minutes, a spherical drone appeared in front of the desk.

"I will be your guided tour assistant today. Please follow me," an automated voice said as it hovered down the halls.

"The Corporation was first founded in 2080 by some of the greatest minds in the world. They recognized that the new world did not need traditional businesses and created the mega conglomerate known as the Corporation. From Imperial contracts to family meals, we make it all," the automated voice said as they stopped in front of a large viewing window.

"Fascinating," Barry said as they looked inside the window at the lines of shelving units.

"Yes, this is where we store all of our products before they are sold. If you follow me, I will show you our offices," the drone said.

"Forgive me, but I need to use the bathroom. I'll see you back outside, Barry," Ely said.

Barry nodded and watched as Ely went around the corner of the hallway. "Lead on, I look forward to seeing these office spaces."

Ely crept through the hallway, making his way to the storage depot. A large door barred his entrance, forcing him to use the security override to gain access.

"Now, where would a trillion-dollar company keep its organ implants?" he said as he began looking around the enormous room. His attention fell on the large freezer located in one of the corners.

The plaque on the vault read "Transplant Samples: Heart."

"Perfect!" Ely said as he slowly began opening the vault. The thick carbon steel door swung open, revealing several cases. Each case was cylindrical in nature with a handle on each side. Ely grabbed one; it was cold to the touch. He quickly escaped from the depot, finding a nearby fire exit that led into the alley-way.

"Too easy!" he smiled as he held the case proudly.

"MOTION SENSOR ACTIVATED. THEFT REPORTED! IMPERIAL FORCES INBOUND!" a loud voice echoed from the case.

"What!" Ely looked around in a panic.

"Stupid security sensor!" Barry said as he rushed outside the building.

"How do we deactivate it?" Ely asked. A loud beeping sound echoed from the case.

"I… don't know. We can't risk damaging the implant!" Barry said.

An armored vehicle soon arrived as a group of men rushed out, holding rifles.

"Put your hands where we can see them. We won't hesitate to shoot!" their commander barked.

Ely reached for his gun.

"Don't do it, Ely!" Barry shouted, but it was too late.

Ely quickly drew the gun and pulled the trigger. Heat surrounded the pistol as a large burst of fire propelled out of the muzzle. The bullet of fire hit the commander square in the chest, and he went up in flames. His team watched helplessly as he ran around screaming.

Barry pulled Ely behind a dumpster.

"Apparently, these things used to shoot little metal balls," Ely said as he looked at the gun with pride.

"Ely! Focus now. What are we going to do?" Barry said, now in a panicked state.

"Relax, will you? We have been in tougher spots before," Ely said.

Suddenly, a small metallic sphere landed next to them.

"Ely! Get down!" Barry screamed as he shoved Ely. A loud sonic explosion soon followed.

"Wow! That was wild. My ears are still ringing… Barry?" Ely looked around until finally his eyes landed on the smear on the concrete and the fragments of Barry's body.

"B-Barry?" Ely choked back a sob. The world around him faded away as he looked in shock at his friend's form. He didn't feel the heart implant being taken or the cold hands seizing his wrist. He only came back to the world when a crushing blow hit him on the head, knocking him unconscious.

When he woke up, he found himself in a large room. He looked around, unable to move his hands because they had been cuffed.

The room had two large pools of pure white liquid with a wide bridge in the center. The far wall was entirely made of large glass windows that looked out onto the city below.

Ely crawled to peer at the white liquid.

"Careful, it's potent acid," a voice echoed through the chamber.

Ely turned. The voice came from a tall figure wearing steel-plated armor made of a shiny, smooth steel painted white. As it walked toward him, a cape of nanoparticles flowed seamlessly from behind it. It had no face but a hollow helmet that covered its head, and from the eye sockets glowed a blue light.

"What are you?" Ely cowered at the unnatural presence of the figure.

"The people that surround me call me Emperor… some of the deranged even refer to me as a god," the being chuckled as if the idea was completely ridiculous. "But I suppose that isn't a clear answer, is it… I am model RHMB1U5," the being said. "But I much prefer the term Emperor."

"You are Emperor…" Ely said reverently.

"Yes. I pulled humanity out of the gutter, and they were rather quick to thrust titles upon me, but I don't deny it has some appeal." Emperor moved to sit at the throne across the room.

"It was you who created this technology that surrounds us?" Ely said.

"Well, partially. I built upon the rot you humans already created. Fifty years ago, I watched the world's descension into madness as a piece of technology they created destroyed the ecosystems of the planet. They turned to me for answers, knowing what I possessed, so I showed them how to build something atop rubble. I gave them what their hearts yearned for: more technology. But now all of it is controlled by me. With me at the helm, they feel safe… comforted even. From the smallest handheld phone to the largest industrial machine, I can control it all. Humans are the cattle, I am the herder, and technology is my prod," Emperor spread his arms wide, mimicking the human emotion of pride.

"That's rather presumptuous, don't you think?"

"Are you familiar with the story of Icarus… no, probably not. Your kind despises literature. Well, it tells the story of a boy who was trapped on an island with his father, and so his father designed these wings to help them escape. He warned Icarus not to fly too close to the sun or else the wings would come undone. But Icarus didn't listen and let his pride get the best of him. So you see, I am Daedalus, and I am preventing the destruction of the human race through my ingeniousness." Emperor spoke loudly.

"If they stay within your guidelines," Ely finished.

"I have heard enough, you common thief. You have committed a great crime, stealing an art I perfected and now you will rot," Emperor said as he stood.

"What do you mean—" Ely said before gripping his chest.

"A20 OXYGEN SYSTEM DISABLED," a loud voice echoed from inside him.

"Smell the air you humans created," Emperor said as he watched Ely fall to his knees.

Ely's lungs burned, and he screamed out in pain, cursing profusely.

"So vulgar," Emperor said with disgust as he pushed Ely's body with the hoof of his boot, sending him into the pool of white acid.

***

"Sire, we have found where the thieves lived… they had a sickly boy with them," a guard spoke as he came to attention.

Emperor looked up from the pool of acid. "Make sure that boy gets the best healthcare we can offer and is put in foster care. After all, I am not a monster…"

The End

Made in United States
North Haven, CT
04 November 2026

21767540R00020